Cambridge Primary

# Hodder Cambridge Primary
# Science
## Activity Book
## A

Rosemary Feasey

## Foundation Stage

**HODDER**
EDUCATION
AN HACHETTE UK COMPANY

The author and publishers would like to thank Chris Lawson, Science and Early Years Lead, Laurel Avenue Primary School, for her support in planning this material.

Although every effort has been made to ensure that website addresses are correct at time of going to press, Hodder Education cannot be held responsible for the content of any website mentioned in this book. It is sometimes possible to find a relocated web page by typing in the address of the home page for a website in the URL window of your browser.

Hachette UK's policy is to use papers that are natural, renewable and recyclable products and made from wood grown in well-managed forests and other controlled sources. The logging and manufacturing processes are expected to conform to the environmental regulations of the country of origin.

Orders: please contact Hachette UK Distribution, Hely Hutchinson Centre, Milton Road, Didcot, Oxfordshire, OX11 7HH. Telephone: +44 (0)1235 827827. Email: education@hachette.co.uk. Lines are open from 9 a.m. to 5 p.m., Monday to Friday. You can also order through our website: www.hoddereducation.com

© Rosemary Feasey 2018

Published by Hodder Education

An Hachette UK Company

Carmelite House, 50 Victoria Embankment, London EC4Y 0DZ

Impression number    10  9  8

Year                 2027  2026  2025  2024

Cover illustration by Steve Evans

Illustrations by Vian Oelofsen

Typeset in FS Albert 17 pt by Lizette Watkiss

Printed in the United Kingdom

A catalogue record for this title is available from the British Library

978 1 5104 4860 5

# Contents

# All about me

☆ Draw the rest of this body. Join the words to the body.

hair

head

arm

hand

leg

foot

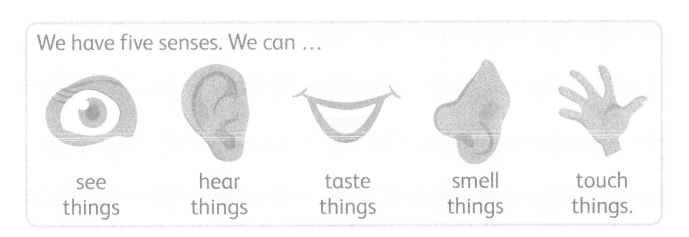

We have five senses. We can …

see things   hear things   taste things   smell things   touch things.

⭐ Draw yourself. Draw something you can see, hear, smell, taste and touch in the circles.

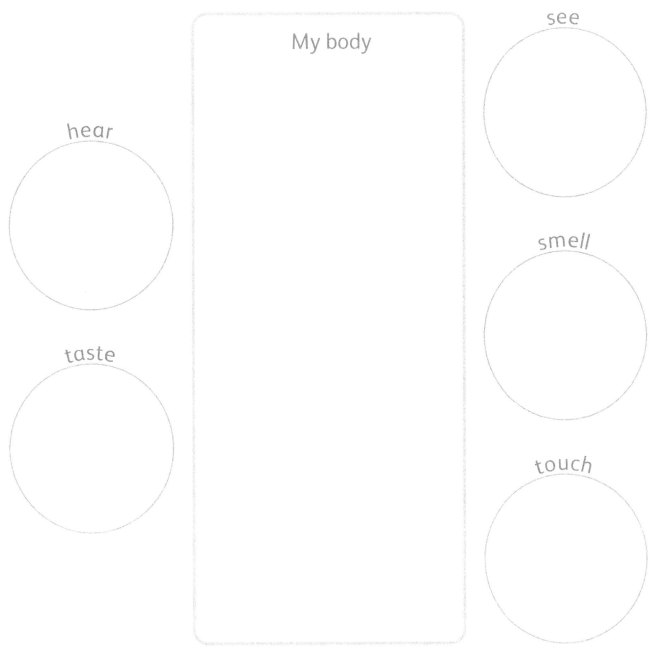

My body

see

hear

smell

taste

touch

# Similar and different

Animals are living things. They can:

| breathe | move | grow | eat and drink |

Humans are animals.

⭐ Aadi and the bird both have a neck.
Circle other parts of the body that are similar.

Aadi                    bird

⭐ Draw a ✗ on parts of the body that only the bird has.

⭐ Circle the parts of the body on the baby giraffe that are different to the girl's body.

Zara

giraffe

⭐ Giraffes are the tallest animals in the world.
Who is the tallest learner in your class? How could you find out?

# What animals need to live

⭐ Draw a picture of yourself in the frame. What do you need to stay alive? Draw one thing in each circle. Use the words to help.

air

sleep

food and water

shelter

⭐ Does a parrot need the same things as you to stay alive?
Draw one thing in each square that a parrot needs to stay alive.

# Animals and their babies

⭐ Join each parent to its baby.

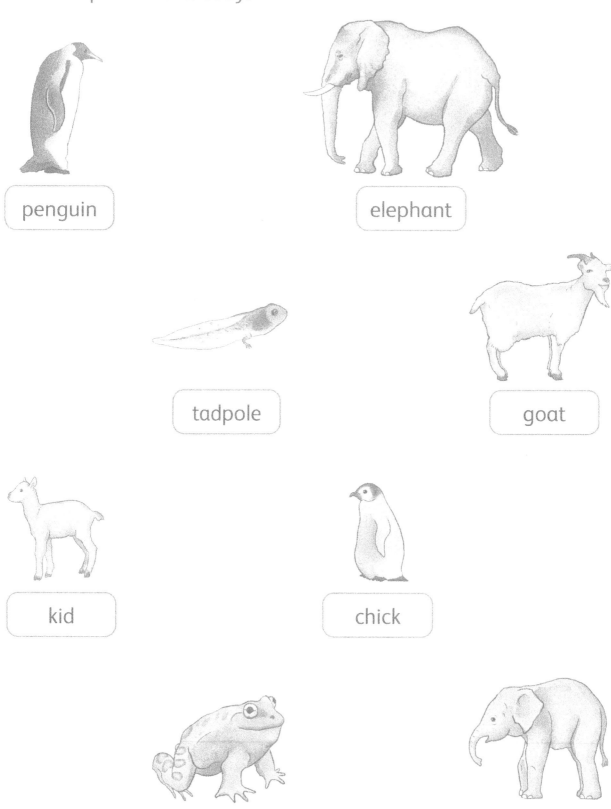

penguin

elephant

tadpole

goat

kid

chick

frog

calf

Which baby animals hatch from eggs?
Write the word **egg** next to the animals that hatch from eggs.

duck _____

cat _____

deer _____

turtle _____

crocodile _____

elephant _____

 Draw another animal that
has babies from eggs.

Write its name.

_____

Its babies are called

_____

# Animals moving

⭐ How do these animals move? Join them to the correct movement.

swim

walk

fly

⭐ Choose another animal. Draw your animal and write how it moves.

My animal is called _____ . It can _____ .

## Animals hiding

A leaf insect looks the same shape and colour as a leaf to blend into its surroundings. This is called camouflage.

 Draw an animal.
Use crayons to colour in and camouflage your animal.

# Where do animals live?

Lions like to live on dry and hot grassy plains. The grassy plain is their habitat.

⭐ Find out which animals like to live in these habitats. Draw an animal in each habitat.

sea

desert

Arctic

grassy plain

Join each animal to its habitat.

| woodlouse | frog | owl |

| pond | trees | logs |

 What is your habitat? Draw a picture of it.

# A habitat near me

⭐ Draw a habitat in your country. Draw animals that live in this habitat.

⭐ Write the name of your habitat. Write the animals in your habitat.

My habitat is _____.

These animals live in my habitat _____

_____.

Make a model of a bird's nest or a spider's web.

What has the bird used to make this nest? What will you use?

How has the spider built this web? What will you use?

 Take a photo of your model and stick it here.
Write what you made it from.

My model is made from _____ .

# The sea

⭐ What do you know about the sea?
Draw and write your ideas in this frame.

# Which animals live in the sea?

⭐ Draw a circle round the animals that live in the sea.

⭐ Draw a ✘ on animals that do not live in the sea.

# Sea animals

⭐ These animals live in the sea. Write a name next to each animal.

| puffer fish | shark | crab | starfish | octopus | whale |

| | |
|---|---|
| | |
| | |
| | |
| | |
| | |
| | |

 Here are some sea animals. Draw the rest of each animal's body.
Use these words to finish the sentences.

| shell | fins | rocks | sea |

I am a limpet.

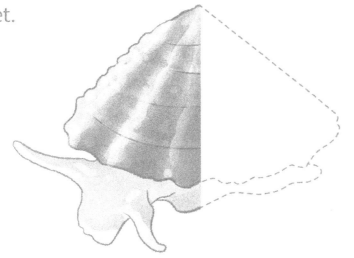

I have a _____ .

I live on _____ .

I am an angelfish.

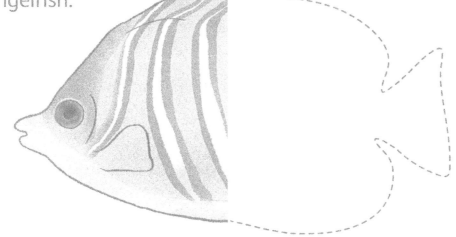

I have _____ .

I swim in the _____ .

# Comparing sea animals

⭐ Look at the squid and the octopus.
How are they similar? How are they different?

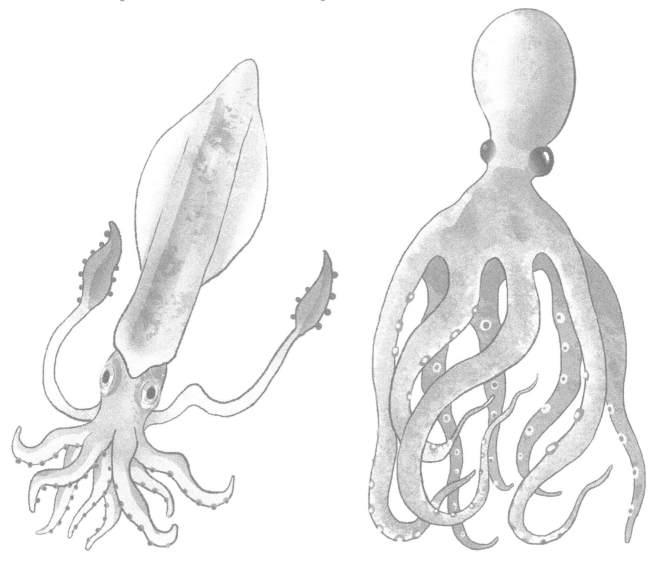

This is a squid.                    This is an octopus.

These things are the same:

_____

These things are different:

_____

# Rock pool habitat

⭐ Which animals and plants live in a rock pool?
Join the animals and plants that live there to the rock pool.

## Sea turtle

⭐ Draw a sea turtle. Label your turtle with these words.

| flippers | shell | tail | head |

⭐ Finish the sentences. Use these words.

| eggs | sea | swim |

Turtles live in the _____.

Turtles can _____.

Turtles lay _____.

☆ Make a model of the life cycle using dough.

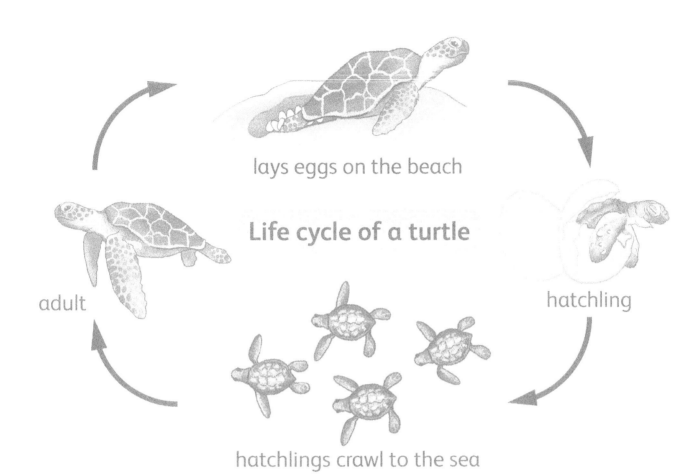

lays eggs on the beach

**Life cycle of a turtle**

adult

hatchling

hatchlings crawl to the sea

☆ Take a photograph of your model. Stick the photograph here.

# Floating and sinking

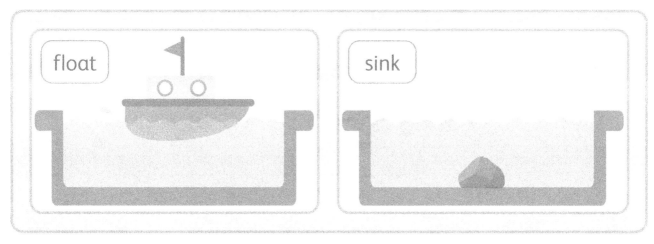

float

sink

☆ Do these things sink or float? Test them in the water tray.

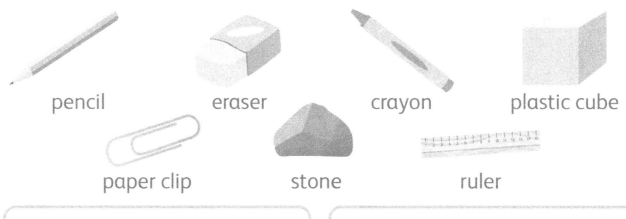

pencil          eraser          crayon          plastic cube

paper clip          stone          ruler

| Draw the things that float. | Draw the things that sink. |
|---|---|
|  |  |

 Collect some other things to test. Do they sink or float?

⭐ Make a boat. Draw a picture of it here.

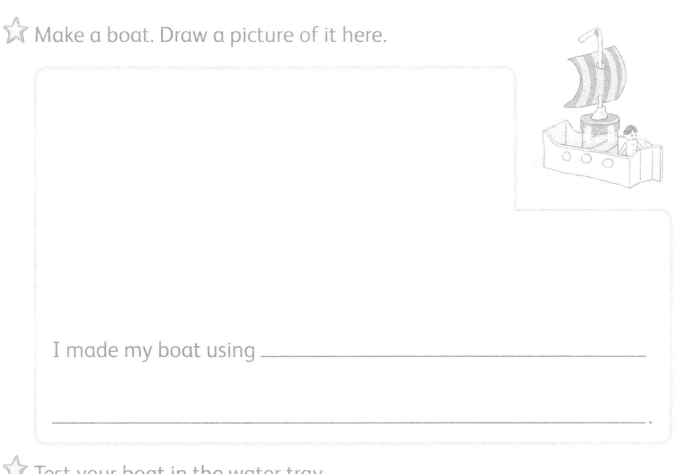

I made my boat using _____

_____

⭐ Test your boat in the water tray.

Can your boat float?        yes ☐        no ☐

How many cubes can your boat carry? ☐

✓ what happens when you put too many cubes on your boat.

It sinks. ☐        It floats. ☐

# Sand

 What does sand feel like? What can you do with sand?

Sand is _____.

Sand can _____.

 Make something using sand. Draw and write about it.

I made _____.

# Looking after the sea

⭐ Circle the things on the beach that could be dangerous to animals and people.

⭐ Where does the litter on the beach come from?

The litter comes from _____.

This litter is not good for animals and people because

_____

⭐ What can we do about the litter? Tell a friend your ideas.

# What can you remember?

⭐ Join the words to the lion's body in the correct place.

nose

eye

mouth

ear

foot

tail

leg

⭐ Which things do both Ishaq and the horse have that are similar?
✓ them.

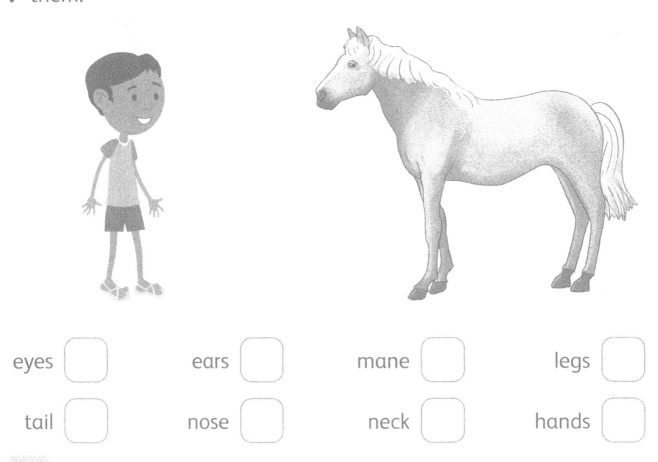

eyes ☐

ears ☐

mane ☐

legs ☐

tail ☐

nose ☐

neck ☐

hands ☐

⭐ Circle the animals that live in the sea.

⭐ Draw three things on top of the water that float.
Draw three things on the bottom of the tank that sink.

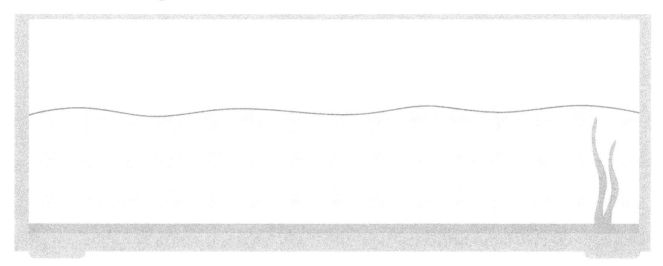

⭐ Here is a beach. Draw three things that could harm the sea turtle.

# Self-assessment

Colour the stars
to show what
you can do!

| | | |
|---|---|---|
| **Animals and us** | I can name parts of my body. | ☆ |
| | I can name the five senses. | ☆ |
| | I can say what I need to stay alive. | ☆ |
| | I can say how I am similar and different to another animal. | ☆ |
| | I can match some baby animals to their parents. | ☆ |
| | I can say what camouflage means. | ☆ |
| | I can say how an animal moves. | ☆ |
| | I can name an animal and its habitat. | ☆ |
| **Above and under the sea** | I can name some animals that live in the sea. | ☆ |
| | I can name parts of the body on a sea animal. | ☆ |
| | I can say how sea animals are similar and different to each other. | ☆ |
| | I can name some animals that live in a rock pool. | ☆ |
| | I can say some things that float. | ☆ |
| | I can say some things that sink. | ☆ |
| | I can talk about the life cycle of a sea turtle. | ☆ |
| | I can name some things that could harm sea animals and people. | ☆ |